This old photograph shows the characteristics of Mendip walls: irregular stone, unrefined technique and no separate cope. The waller is laying a 'binder' (a long stone stretching into the wall) and another can be seen in the exposed section of wall to his right.

Dry Stone Walls

Lawrence Garner

Published in 2012 by Shire Publications Ltd, Midland House, West Way, Botley, Oxford OX2 0PH. Website: www.shirebooks.com

First published 1984; reprinted 1985, 1987, 1988, 1991, 1992, 1995, 1997, 1999, 2001 and 2003. Second edition, updated and with additional illustrations in colour, 2005; reprinted 2007, 2009 and 2012. Shire Library 114. ISBN 978 0 74780 620 2.

British Library Cataloguing in Publication Data: Garner, Lawrence. Dry Stone Walls. – 2nd ed. – (Shire album; 114) 1. Dry stone walls – Great Britain I. Title 631.2'7'0941. ISBN 978 0 7478 0620 2.

Cover: Malham View. Taken in Malhamdale on the southern edge of the Yorkshire Dales, showing the farming landscape with its miles of ancient limestone dry-stone walls.

ACKNOWLEDGEMENTS

Photographs are acknowledged as follows: Dry Stone Walling Association, pages 1, 7, 31 (bottom); Lawrence Garner, pages 3, 4, 5, 12, 13 (both), 14 (both), 15 (both), 16 (both), 17, 18 (both), 20 (bottom), 22, 23 (both), 24 (both), 25 (top), 26 (both), 27 (both), 28, 29 (both), 30 (both), 31 (top), 32-3, 34 (both), 38; Cadbury Lamb, pages 6, 9, 25 (bottom), 35, 36, 37, 39; Paul Stevenson, cover; Judith Weston, pages 19, 20 (top), 21.

Printed in China through Worldprint Ltd.

Contents

A traditional Lake District sheepfold, newly restored.

Old field walls in Ribblesdale; the newly repaired section shows that they still serve a useful purpose.

Why dry stone walls?

A dry stone wall is one built entirely or mainly without the use of mortar; it relies for its strength and durability on the skilful placing of stones so that each one is locked securely in position.

No one who has travelled widely in Britain can have failed to notice the part that dry stone walls play in the agricultural economy of upland areas. Many visitors, however, accept them as a quaint addition to the rural scene without asking what the walls are doing there.

The immediate answer is that on mountains, moorlands and fells vegetation does not grow readily. Trees are scarce and a thick hedge is an impossibility. From the very earliest times walls were an obvious method of enclosure, for stones usually littered the original unimproved pastureland or lay in plentiful supply a few inches beneath the surface.

But walls have more positive advantages. Many of the walls that enclose some of Britain's highest terrain have stood for two hundred years. A good waller today will expect his work to outlive him provided that the owner gives the wall the small amount of maintenance needed. Compare this with a wire and post fence, which not only requires frequent attention but will rust and rot even in favourable weather conditions.

Where hedges are an alternative, maintenance once again becomes a chore if they are not to spread and encroach on useful land, and neither fence nor hedge provides shelter against cold wind and driving rain. At lambing time a wall can be a life-saver to ewes and lambs and a godsend to the shepherd on exposed heights. A final virtue, often overlooked, is that they are fireproof. Walling areas are usually tourist areas too, and fires started carelessly in dry conditions are an increasing problem. Hedges and fences simply provide additional fuel whereas a wall will act as a firebreak.

There are, however, disadvantages. To retain the most intrepid breeds of sheep, a wall needs to be over 5 feet (1.5 metres) high, although a strand of barbed wire along the top of a lower wall will usually foil attempts to leap over. Compared with a fence, a wall takes a long time to erect; a waller on his own will average 5 to 6 yards (4.9 to 5.5 metres) a day. Above all, walls are expensive.

Labour costs for building a new wall or rebuilding an old one vary enormously depending on such factors as location, height, difficulty and competition for the work. If stone is brought in, the cost goes up alarmingly. Yet, considered as a long term investment, a wall is bound to be less expensive than fencing. After 150 years a wall that has received proper attention will still be serving its purpose satisfactorily, having outlasted a dozen fences. However, to the majority of farmers concerned with their cash flow the cost of repairing walls, whether counted in money or time, will be a deterrent and we shall continue to see odd bits of wire slung across collapsed sections of wall.

This leads to the controversial subject of the aesthetic value of dry stone walling. The extent to which farmers and other country dwellers have an obligation to keep rural areas looking attractive is a matter of opinion, but there is no doubt that some of Britain's favourite tourist areas are

A stretch of remote Cumbrian moorland patterned by dry stone walls.

A decorative wall surrounding an information point near Selkirk in the Scottish Borders. The double cope is unusual and perhaps not advisable – the top copestones seem to be resting insecurely on their very uneven base.

those which are poorest economically and need the additional income that visitors bring. Dry stone walls are a tourist attraction not so much in themselves as in the way they enhance the landscape.

Think of the Cotswolds without their golden stone or the Peak District of Derbyshire without its white limestone walls. Walls impose a natural pattern and give a sense of scale to an otherwise featureless prospect, without being assertive or detracting from the beauty of the landscape.

There are encouraging signs that the visual value of walls is increasingly being considered along with their agricultural value. After all, nobody travels a long way to gaze at wire fences.

In this remote area of the south Pennines the old drovers' roads have been walled to prevent sheep straying on to enclosed land. Stones still litter the fields, and the limestone outcrop would provide a further source of raw material.

The development of walls

A full history of dry stone walling would be wide-ranging in space and time. It would need to consider structures in Britain that are thousands of years old and make Hadrian's Wall a recent development by comparison. This brief survey can only outline the history of the walls that are easily visible today.

The most frequently observed pattern of walls in upland areas shows three distinct stages. Around the isolated villages and homesteads the enclosures are small and irregular. Beyond these, but still close to home, are larger fields with straighter walls forming a roughly rectangular pattern, often long and narrow. Further out still, on the highest ground, the walls run straight as a ruler for miles, enclosing very large tracts of land.

The homestead walls, often arbitrary in direction and over-massive for the modest fields they surround, are among the very oldest still standing. Some of them date from the fifteenth century and represent early attempts at progressive individual farming after the communal methods of an earlier age.

The farmers of the time still relied on common land for most of their grazing, but enclosure gave them the chance to grow feed, to check and treat livestock and, very important, to confine their stock at intervals in order to manure the land. The walls built were primitive in technique, and their irregular lines were due to the whim of the owner, the need to avoid obstacles or perhaps a wish to incorporate boulders that were too big to be moved easily.

People studying these walls for the first time are inclined to ask why anyone bothered to wall in such tiny fields. One answer is that these early subsistence farmers worked on a miniature scale, but it may be equally true that repeated ploughing threw up immense quantities of stone that had to be methodically cleared. Building walls was a useful form of waste disposal.

The sixteenth-century boom in wool production was an incentive for farmers to increase and improve their flocks. Larger enclosures were required and, with new statutory backing, the tendency was to take in more land close to the villages. This was a continuing process during the sixteenth and seventeenth centuries, usually after communal deliberations and agreements. The result was often a radial pattern of narrow, elongated fields stretching out into the nearer common land.

This second stage of enclosure led in some areas to the use of walls for a different purpose. As land became an increasingly valuable commodity, it became necessary for the first time to define the extent of each settlement's common land, and some very long walls were built for this purpose by communal labour.

However, this activity was on a small scale compared with the great period of wall building that started in the late eighteenth century and continued into the early nineteenth. Both grazing and arable land was affected by the development of farm machinery that needed large fields for optimum use, and by a sharply increasing demand for food and clothing from the growing industrial towns. For the first time scientific cattle breeding and arable cultivation became an occupation for gentlemen, who took full advantage of their property rights in the great enclosure movement.

This was the golden age of the professional wallers, sometimes local and sometimes nomadic, who worked to surveyors' specifications and built the thousands of miles of walls across former common land and virgin terrain in some of the remotest and most inhospitable regions of Britain.

For the first time walling became recognised as a specialist craft and sophisticated techniques were established to make it possible to build sound walls at considerable speed. The methods developed in the eighteenth century remain virtually unchanged today.

This account is greatly simplified but the pattern of events was remarkably similar throughout the major walling areas of south Scotland, the Lake District, the Cotswolds and the large area of the Pennines stretching from the Tyne down to north Staffordshire. In the more sparsely populated areas like Wales and south-west England there was never the same intensive activity; long walls were built to define grazing districts, but seldom by professionals and never with the urgency and competitiveness of more prosperous regions.

By 1850 virtually all the agricultural walls had been built. The vast majority of those standing today originated before the middle of the nineteenth century. In remoter places, where they are free from interference,

A pleasant piece of decorative walling at Rosemoor, the Royal Horticultural Society's garden near Torrington in Devon. Children find this sort of flat top irresistible, so secure mortaring is essential.

they have remained untouched; elsewhere piecemeal repair and renovation have taken place. Unless accurate documentation is available it is very difficult to date the construction of any wall since the same stones and the same techniques will have been used in renovation.

Anyone wishing to see a potted history of dry stone walling should go to Malham in North Yorkshire. In the course of a two-hour walk you can see the irregular homestead walls, the early 'intakes', the majestic eighteenth-century walls, the limestone pavements sliced away by the builders and, on the highest ground, large tracts of land in its original, stone-littered state.

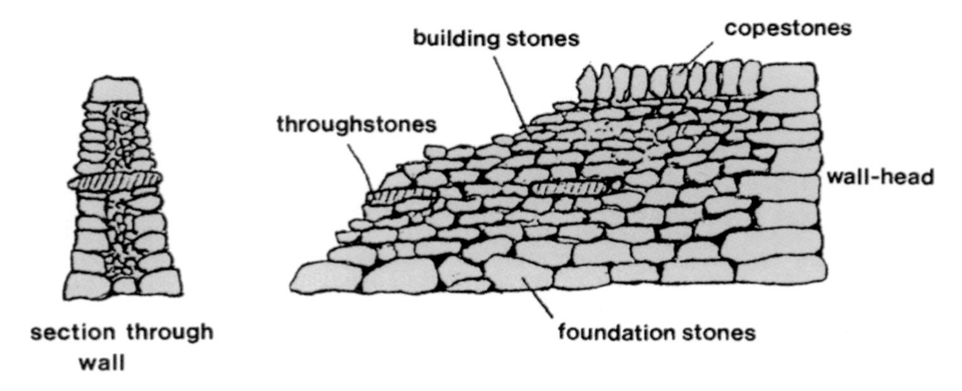

The elements in the construction of a wall.

The construction of a dry stone wall

Intelligent appreciation of dry stone walls must begin with a knowledge of the principles involved in building them. Regional variations are discussed later; what follows here is a description of a wall built to a specification which would be generally accepted as adequate for strength and durability.

The front view and cross-section in the drawing above show the various elements of the wall. It consists of two sides tapering towards the top, with the cavity between the sides filled with small stones and with *throughstones* at regular intervals to tie the sides together.

The first stage is the preparation of the foundation bed. If the wall is to be of the standard height of 4 feet 6 inches (1.37 metres), the foundation will be between 28 and 32 inches (710 and 810 mm) wide. (Dimensions are the result of traditional wisdom rather than science and it is surprising to find that similar optimum measurements were arrived at independently in widely separated areas of Britain.)

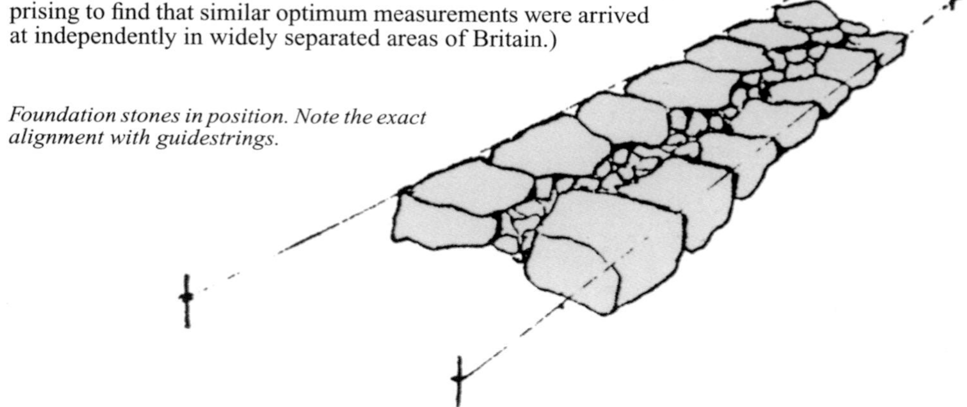

Foundation stones in position. Note the exact alignment with guidestrings.

The waller pegs out guidestrings along the proposed line of the wall at a width of about 3 feet (910 mm). Within these strings he excavates the soil until he reaches firm subsoil – usually not much more than 4 to 6 inches (100 to 150 mm) down in upland areas. Projecting stones are prised out and the length is levelled off to provide a firm earth bed. The guidestrings are then moved in to the specified foundation width and the laying of the *footings* begins.

The foundation stones will be the biggest and squarest in the heap. They are placed in a double line with their widest and flattest side downwards to prevent their being driven into the earth by the weight of the stones on top. The space between the two rows is carefully packed with small stones known as the *fill* or *hearting*. The fill will be continued to the top of the wall to stop the two sides collapsing inwards.

The aim now is to build successive courses so that the two sides of the wall taper to a width at the top approximately half that of the base. The degree of taper is known as the *batter.* Some wallers claim to be able to judge the batter by eye, but most professionals will use a *batter frame* made to the dimensions of the wall. With one of these frames at each end of the length and strings tied between them, the correct batter at any stage is accurately indicated.

A beginner may want to move the strings up the frame after each course, but an experienced waller will start with them set several courses up. When the stones are being laid, the batter is achieved by setting each new course in a little from the last.

In laying courses on the foundations the waller will be following certain basic principles. The biggest stones will be used first, shared out equally on both sides. The space between the stones in one course will be firmly

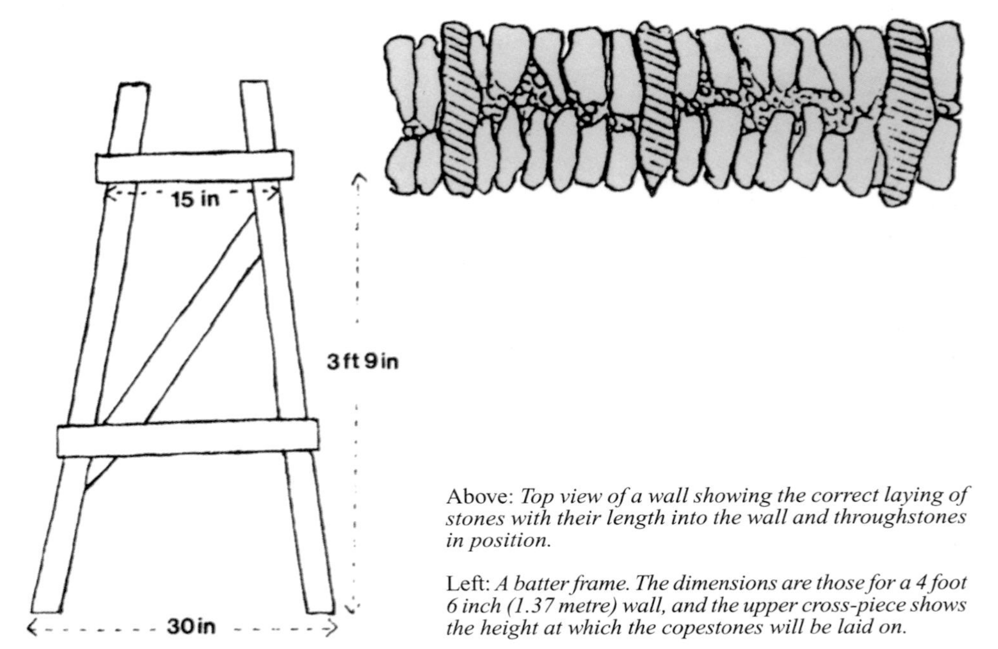

Above: *Top view of a wall showing the correct laying of stones with their length into the wall and throughstones in position.*

Left: *A batter frame. The dimensions are those for a 4 foot 6 inch (1.37 metre) wall, and the upper cross-piece shows the height at which the copestones will be laid on.*

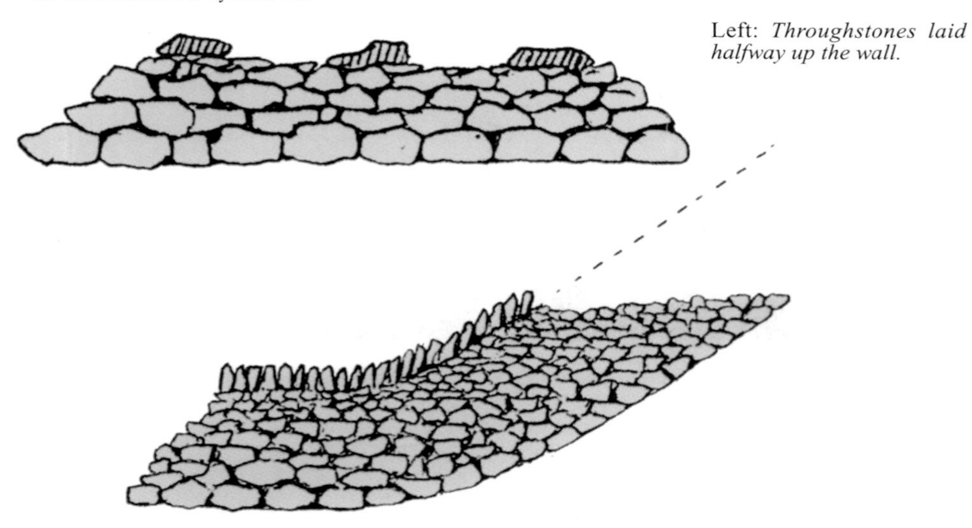

Left: *Throughstones laid halfway up the wall.*

Building on a slope. The courses are laid horizontally, while the height of the wall (shown by the broken line) is accurately maintained. The copestones lean slightly uphill.

covered by stones in the course above, as in bricklaying. This is known as *breaking* or *covering* the joints. Where stones have a definite length they will be placed with the length running into the wall and not along it. In this way they will be locked firmly into place. Where possible, the flat or vertical end of the stone is placed to the outside, and the stone is either laid flat or sloping slightly downwards; it should never tilt back towards

With the foundation laid, the batter frame is positioned. This will ensure the correct taper for the wall.

The guide-strings are stretched between the two batter frames and the wall begins to rise.

the centre as this will funnel rainwater into the middle of the wall. The two sides of the wall are built up together and the filling is continued scrupulously as the outer stones are laid.

At a specified height (in this case halfway up the double wall) the two sides will be levelled off and the throughstones laid on. *Throughs* are long stones stretching across the two sides and making good contact with each. Their purpose is to act as ties to hold the sides together and stop the wall bellying out. The more throughs the wall contains the stronger it

The wall is levelled off ready for the large flat stones (cover-bands) on which the cope will rest.

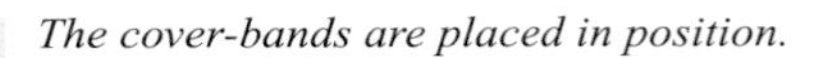

The cover-bands are placed in position.

Below: *Large boulders forming the cope complete the wall. (Notice the heavy throughstone halfway up the wall.)*

A fine piece of Lakeland wall levelled and ready for coping. The material is slate.

Wallers finishing a Yorkshire competition length. Note the two rows of projecting throughstones.

The two varieties of Derbyshire walling stone – white limestone and dark millstone grit – are illustrated at the National Stone Centre, Wirksworth, Derbyshire. Between them is a 'squeeze stile'.

will be, but in some areas they are hard to come by. Ideally there should be throughs at intervals of not more than a yard.

The guidestrings will now be moved up the frame to the height at which the copestones are to be put on, in this case at about 3 feet 9 inches (1.14 metres). The building continues with the stones getting smaller and needing even more careful laying to lock them into place. When the wall once again reaches the strings its width will be 14 to 16 inches (360 to 410 mm), and it will be levelled off again to receive the copestones.

Copestones have two functions apart from their decorative effect. Firstly, they add considerable weight to the wall, weight which is specially vital to stabilise the lighter stones in the upper half. Secondly, they act as throughstones and must make good contact with both sides. The ideal copestone is fairly thin and flat and is placed upright on the wall. In this way more stones, and therefore more weight, can be placed in a given length. Copestones should add about 9 inches (230 mm) to the height, bringing the wall up to the specified 4 feet 6 inches (1.37 metres).

A wall-head marking a division of ownership. Note the long stones used to reinforce it.

A very solid stile demonstrated at the National Stone Centre, Wirksworth, Derbyshire.

SPECIAL FEATURES

It is sometimes necessary to incorporate special features for which particular techniques are required.

Where a wall comes to a free-standing end, for example at a gate, extra stability has to be built in to make it self-supporting. It is achieved by alternating throughstones with long stones that stretch back into the wall. This is known as a *gate-end, cheek* or *wall-head.* Two wall-heads butted together without forming a break in the wall are often used to mark a change of ownership.

Where a wall crosses a footpath a stile must be introduced to avoid disturbing the copestones. Where cattle are enclosed this can take the form of a V-shaped opening in the upper half of the wall, called a *squeeze stile* for obvious reasons. The wall on each side of the gap must be strongly reinforced, as for a wall-head. Where no opening can be permitted the usual method is to lay one or more extra long stones across the interior of the wall to provide a projecting foothold on each side. On a public footpath as many as three such stones will be used, staggered to provide an easy 'staircase' on each side.

One of the commonest features is a small hole in the bottom of the wall to allow sheep access from one field to another. The most important component is a suitably stout stone to act as a lintel across the top. There are many regional names for these holes, such as *creep-holes, cripples, hogg-holes, lunkies* and *smoots*. The same technique can be used where a wall has to cross a stream.

A sight that never fails to impress tourists is a wall built up a very steep

This wall near Hawes, North Yorkshire, with its creep-hole for sheep, is a good example of the primitive style that preceded the enclosure walls built by professionals.

gradient, although building it is not difficult provided that the courses are kept horizontal. The tricky part is placing the copestones, which most wallers agree should lean uphill so that if one is pushed out the remainder do not collapse like dominoes.

This Welsh wall is built down to the edge of a precipice and shows an unconventional but very skilful technique. The long slabs lock in securely to support the end, and one of them is used ingeniously to hold up the copestones. Horizontal coursing is maintained throughout.

Regional stones and styles

Although the basic principles of dry stone walling remain the same everywhere, regional variations arise for two reasons – the characteristics of the local stone and the prevailing agricultural conditions.

When the walls were originally built the stone was found either by clearing it off the land or by digging for it nearby, so wallers worked with whatever material was to hand and adapted their methods according to its characteristics. If throughstones were not available some other way had to be found of making sure that the wall did not fall apart. If the stones were small and light then high walls could not be built.

Walls enclosing cattle were built differently from walls enclosing sheep, and styles would vary according to the breed of sheep. Within these limitations wallers would develop their personal styles. The major regional variations to be seen in the walling areas of Britain will be described in this chapter.

SCOTLAND

In Scotland dry stone walls are called *drystane dykes* and two areas in particular, the south-west and the north-east, have been influential in establishing building styles.

In the south-west the Galloway dykers seem to have perfected techniques from the eighteenth century onwards to deal with a wide variety of stone. In the standard double dyke, which is 4 feet 6 inches (1.37 metres) high, the stones are laid as far as possible in level courses, whatever their shape. *Throughbands* are placed halfway, although in a *march dyke* (boundary wall), where the height may be 5 feet (1.50 metres) or more,

Heavyweight dry stone walling used to shore up a landslide hazard in the Scottish Highlands.

In this Ross-shire wall the large boulders not only stabilise the smaller stones beneath but raise the height of the wall by one third. They also resist the attentions of cattle.

A demonstration length by a Sutherland dyker at the National Stone Centre. 'Single dyking' – the piling of large stones with no lateral support – is a difficult technique.

A textbook example of Scottish walling – effective distribution of stones, precise levelling off and copestones of uniform size.

two rows are usually found. A distinctive feature of Galloway double dykes is the *cover-bands,* stones laid right across the top of the dyke underneath the cope and projecting slightly on each side.

Great stress is laid on careful hearting or filling and on laying stones with their length into the wall. Copestones will be vertical or slightly inclined and they are often *locked* (wedged with small stones) to hold them firm.

A good deal of single dyking is found in south and west Scotland. Instead of two walls tied together and filled, the structure consists simply of substantial stones piled one on another. Since the stones are seldom square this is a highly skilled operation. There is no hearting, so daylight shows through the wall – a feature which is said to deter sheep from jumping the dyke because it looks unsafe.

The true Galloway dyke is a combination of both methods. The lower section is built in conventional double style, cover-bands are laid on, and the rest of the dyke consists of single boulders. It is an illustration of the Galloway dyker's need to deal with a wide variety of stone.

In the north-east of Scotland the predominant stone is granite and this has produced a style of building with large blocks, the weight of which will keep a dyke stable without the use of through bands. Since the dykes often have to contain beef cattle projections are avoided as they form convenient 'scratching points', where stones will quickly be dislodged. Granite blocks tend to be regular in size and shape and can be laid in very straight courses, with smaller stones filling any interstices. The familiar thin, vertical copestones are seldom seen since they would add little appreciable weight. Where a cope is used it is formed of massive stones weighing anything up to a hundredweight (50 kg) each. Granite dyking makes great demands on physical strength.

WALES

Welsh walls cannot match the conscious craftsmanship of the Scottish dykes, probably because there was never the same tradition of professional walling. The enclosure movement was not so strong in Wales, and most of the walls on hill farms are very old.

Limestone, sandstone and gritstone are the prevailing materials in South Wales and their use is often more optimistic and instinctive than skilful. The merits of throughstones are frequently ignored, so the commonest fate of the walls is a gradual bellying-out in the middle. The casual use of copestones, or their omission, is a further weakness. The gritstone of the coalfield areas provides more regular stones and walls of this material seem to survive longer. However, renovation work on sounder principles produces respectable results, although Wales suffers from a lack of professionals even today.

The mountain walls of Gwynedd are among the most impressive to be found anywhere. Many of them run for miles over very severe country, enclosing vast grazing areas and often climbing slopes of more than 45 degrees. They comprise a mixture of excavated shale of a very fissile kind and hard volcanic rock cleared from the land. Unlike the English enclosure walls, they seldom run straight. Deviations were made to avoid the worst obstacles and to take advantage of useful outcrops and embedded boulders.

A consumption wall near Harlech, Gwynedd. About 6 feet (1.8 metres) thick, it served as a dump for stones cleared off the fields.

An elegant slate-built folly with a domed roof near Beddgelert, Gwynedd.

The use of slate as a walling material is a feature of the quarrying districts, and fences made up of thin slate slabs bedded in the ground and held together with a single strand of wire are common.

In the most popular tourist areas copestones are often cemented. Elsewhere they tend to be omitted if the building stone is heavy enough; otherwise they are functional with little attempt at decorative appearance.

THE LAKE DISTRICT

The geology of the Lake District is extremely complex and the wide variety of stone leads to some varied styles of walling.

The south-east area between Lancaster and Kendal is dominated by carboniferous limestone, giving walls of whitish-grey stones, small and irregular in shape. They do not respond well to hammering and considerable skill is needed to bind them closely and firmly. Further north, in a broad band between Kendal and Ambleside, the walls tend to be made up of small flat slabs, which are easier to lay than the

A well-crafted urban wall of local slate at Bowness-on-Windermere, Cumbria.

A demonstration wall of dressed Cumbrian sandstone with step stile.

irregular limestone but which require a good deal of patience, since the walls are slow to rise. As with Welsh slate, larger slabs are sometimes formed into a stone fence.

North of Ambleside and stretching up to Keswick is an extensive area of volcanic rock. It appears in a variety of forms, including the famous 'green slates', and is much harder than the stone mentioned so far. It does not break down easily; the walls built with it look lumpy, with stones of irregular size, and often very large boulders are incorporated at the base.

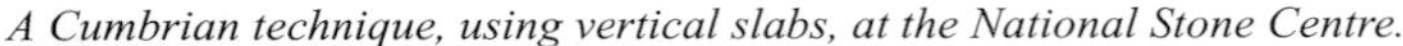

A Cumbrian technique, using vertical slabs, at the National Stone Centre.

A big Cumbrian barn now converted to a house. The throughstones are prominent.

Similar in character are the granite walls in a small area to the west of this geological band.

Finally, in the extreme north of Lakeland there are the Skiddaw slates. They are fissile and cleave easily, forming walls that are dark in colour and regularly coursed.

Lake District walls are conventional in size and building method, with at least one row of throughstones, and often two in higher walls, and are coped with flat, vertical *cams*. A journey up the M6 motorway will reveal most of the styles described here.

'Cock and hen' coping in Cumbria, slanting in traditional Lakeland fashion. It is far from elegant, but the thin, sharp slabs are a deterrent to agile sheep (and possibly humans too). This example is at Brantwood, the former home of John Ruskin near Coniston in Cumbria.

THE PENNINES

'The Pennines' refers here to the vast walling area stretching from the Tyne down to north Staffordshire. Two kinds of stone predominate, carboniferous limestone and the much darker millstone grit found mainly in South and West Yorkshire and in parts of Derbyshire.

The limestone walls are near white in colour where they have not been affected by years of industrial pollution. They are built of fairly small, irregular stones, normally used with little attempt at coursing or shaping, for Pennine wallers rarely use a hammer.

In some limestone areas throughstones are not easily found, in which case the walls need to be wide-based. Various attempts were made in the

Classic Yorkshire professional style – two rows of throughstones and a band under the cope.

This Yorkshire wall has one band of throughstones near the top, to reinforce the weaker small stones.

The National Park Centre at Hawes, North Yorkshire: a pleasant piece of architectural walling in the form of a spiral.

A pattern of enclosure walls on improved land at Kettlewell, North Yorkshire.

An unusual example of buttressing on a Derbyshire railway embankment.

past to rectify the deficiency. There is evidence that long stones were carried considerable distances for the benefit of the professional wallers, who did not fail to take advantage also of limestone outcrops and pavements from which slabs could be sliced.

The gritstone walls are a complete contrast. The dark grey-brown stone is 'blocky' and can be laid in regular courses. Throughstones are plentiful, so gritstone walls are often narrower as a result. The surface of the stone is rough and provides excellent adhesion. The neat effect of gritstone sometimes inspired the builders to shape the copestones carefully, with elegant results in some villages. In remoter parts copes, both in gritstone and limestone, tend to be rough but functional.

THE COTSWOLDS AND THE SOUTH-WEST

It is ironic that the mellow stone walls of the Cotswolds, admired by thousands of tourists each year, are regarded by experts from other parts of Britain as inferior.

Cotswold wallers have many difficulties to contend with. The distinctive Jurassic limestone tends to be soft and vulnerable to frost and road salt. Building stones are mainly small and light, making the finished wall less stable, and it is difficult to find copestones (known as *combers* in this region) big and heavy enough to function effectively. Throughstones are often lacking. Even so it is difficult to defend the Cotswold waller's inclination to lay stones with their length along the wall rather than into it.

Cotswold technique is specialised. The frost danger requires that stones should be inclined slightly downwards towards the outside in order to shed water. Unlike their northern counterparts, wallers here use a hammer

An old Cotswold estate wall coped with heavy flat stones. Traditional rough copestones were not considered suitable for parkland.

continually to shape the soft stone for securer fit or to improve the face. This can lead to exceptionally neat walls of almost brick-like regularity. The height is about a foot (300 mm) lower than that of northern walls, partly because of the limitations of the stone and partly because traditional Cotswold breeds of sheep rarely tried to climb over a wall.

The lack of suitable copestones has meant that walls that are subject to traffic vibration or the attentions of the public have acquired mortared tops, which is regrettable but understandable in the circumstances. A more recent tendency is the pouring of mortar into the middle to secure the building stones; the effects of this in the long term have still to be discovered.

In the Mendip area there are similar limitations. Carboniferous limestone in its usual

A fine piece of new Cotswold walling forming the curved entrance to a housing development. Note the square-cut cope; new Cotswold stone is easily shaped.

An old Cotswold wall still in excellent repair, owing to the judicious distribution of large blocks to hold smaller and lighter stones in place.

small, irregular form predominates and the problem of insufficient throughstones is surmounted to some extent by the use of longer stones that extend about two-thirds of the way across the wall. As in the Cotswolds, it is the practice to lay stones with their length along the wall. Walls without a separate cope are quite common and there is often very little batter or taper.

A wide variety of stone enclosures is to be found in Devon and Cornwall. Freestanding dry stone walls of the conventional kind are to be seen on the moorlands such as Dartmoor, where granite provides substantial stones. Copes are usually of the vertical type but occasionally turf is used.

However, the characteristic feature of the south-western peninsula is the stone hedge in its various forms, basically an earth bank faced with stone and topped with turf, brushwood, stone or a permutation of these. Typically the base and height measurements are the same and the hedge tapers to rather less than half the width

A new solution to an old problem. This wall under construction at Northleach, Gloucestershire, has its base stabilised by concrete blocks running through its centre.

A rare sight: a rebuilt Cotswold field wall with a shaped cope. The supplementary fencing becomes necessary if agile breeds of sheep are kept.

of the foundation. The faces of the hedge are slightly concave.

The way the facing stones are laid depends on the material available. 'Slab' material, such as quarry waste or slate, can only be used vertically with its length into the bank and there are two main ways of laying it – in straightforward courses of upright stones or in a herringbone pattern, with the stones leaning opposite ways in alternate courses. Many wallers in any case use a herringbone method for the top courses in order to provide better rooting for the vegetation on top. The turf or brushwood

A Cornish stone hedge. The herringbone arrangement of stones at the top allows the turf roots to spread down and bind the whole top firmly.

roots bind the top of the hedge very firmly provided that the earth is not washed away prematurely.

Devon wallers using granite or limestone frequently favour a technique called *chip and block*, where gaps caused by the irregularities of the rough stones are filled by smaller *chips*. In roadside or estate hedges the stones may be dressed to achieve a smooth flush finish.

Stone hedging is the only branch of the craft where it is considered correct to lay stones tilting back towards the centre. Where vegetation is relied on to bind the work, rain must be encouraged into the centre.

A stone hedge is most vulnerable just after building. As the earth settles inside the facing it can push the bottom courses outwards so that

A roadside wall of Mourne granite blocks in County Down, Northern Ireland.

bulges develop. The theory behind the concave face is that this process will eventually produce a straight surface. The other main danger is the shifting of stones before rooting has taken place. This can happen because of interference from cattle or people or during unusually prolonged and heavy rain.

NORTHERN IRELAND

The most varied and interesting walls in Northern Ireland are to be found in County Down, and especially in the region of the Mourne Mountains. In remote areas huge boulders cleared from the fields are stacked two deep to make a substantial boundary. Smaller boulders are

A well-constructed wall of heavyweight boulders in County Down.

also used, although considerable skill is needed to build effectively with the smooth, round stones, which do not have much adhesion. In direct contrast are the occasional elegant roadside walls built with neat blocks of Mourne granite. It is also possible to see stone-faced turf banks similar to the traditional Devon hedge. The great attraction for walling enthusiasts, however, is the Mourne Wall, constructed early in the twentieth century to enclose the catchment area of a reservoir system. Forming a 20 mile (32 km) boundary, it is the longest continuous wall in Britain and took over eighteen years to build. Technically it is unremarkable, incorporating whatever stones the builders came across, but the achievement is spectacular. The wall can be seen at the Mourne Reservoirs country park.

The astonishing Mourne Wall, which encloses a reservoir system and is over 20 miles (32 km) long. It took eighteen years to build.

Freshly dug limestone is fairly soft and can be easily dressed. Virtually every stone in this stretch of demonstration wall at Carsington Water, Derbyshire, shows the mark of an axe.

The craft today

Dry stone walling is not a dying craft. Through such organisations as the Dry Stone Walling Association (DSWA), the Rural Development Commission, the British Trust for Conservation Volunteers, the Agricultural Training Board and other training agencies hundreds of people each year get an opportunity to acquire varying degrees of skill and to put that skill into practice. Few new walls are being built – the cost of stone prohibits it – but many miles are being restored as a result of a new sense of the importance of conserving the traditional landscape.

The professionals remain the elite among wallers. Whether employed by contractors or operating their own small businesses, they all have the ability to complete each day a substantial yardage of high-quality work. What distinguishes them from the skilled amateurs is not so much the mastery of the craft as the speed and economy of work that makes walling look deceptively easy.

A significant development has been the establishment by the DSWA of a professional register, compiled by means of examination and assessment of work. The list of registered wallers is now surprisingly long and includes an encouraging number of younger professionals operating in

Wallers at the Highland Folk Park, Newtonmore, Highland, working with an unpromising mixture of boulder and slab. The lower section of the wall in the foreground needs more filling to bring the stones up level.

areas where it was previously difficult to find skilled craftsmen.

Technically, any waller who does not earn his living from the craft is an amateur, but this is a misleading term. In many upland areas walling has always been a skill handed on by farmers to their sons and an enormous amount of repair work is done by farmers, who regard it as routine. The Agricultural Training Boards have done much to foster walling skills in this sphere. Few farmers, however, have time to do more than patch their walls and they cannot contemplate complete renovation. This situation will improve only when realistic grant aid makes it economic for professionals to be employed.

Volunteer workers have become increasingly prominent in dry stone walling. They are usually young, often from urban backgrounds, and will probably sample walling as one of a variety of conservation skills. The principal organising and training agency is the British Trust for Conservation Volunteers, which confines its activities to publicly owned areas and national parks and is careful not to undertake tasks that might remove work from professionals.

A final group, often overlooked, consists of those who tackle do-it-yourself jobs on their own property or find the craft a relaxing hobby. These enthusiasts are often catered for by local branches of the DSWA or by organisations of 'friends' within national parks, and there is no doubt that amateur volunteers, through local initiatives, can make a considerable contribution to the restoration of walls that would otherwise receive no attention.

An equally important trend has been the expanding programme of competitions, demonstrations and other events designed to publicise the craft and bring to public attention the importance of dry stone walls in the landscape. The first organisation dedicated to this task was the Stewartry of Kirkcudbrightshire Drystane Dyking Committee, whose major achievements were the establishment of a biennial competition at Gatehouse of Fleet (still the biggest of its kind in the world) and the founding of the DSWA in 1968.

This pioneer work has been taken up by a growing number of local groups who ensure that the waller's craft is demonstrated at country fairs, agricultural shows and similar occasions.

Thatching and blacksmithing were also once thought to be dying trades. They have survived partly because enough people came to realise, in the nick of time, that traditional methods and materials were more acceptable than modern technology, partly because of a wider concern that an aspect of Britain's heritage should not be lost, but mainly because a new generation of competent craftsmen emerged to supply an increasing demand.

All these factors are now working to ensure the survival and expansion of the craft of dry stone walling.

A dry stone feature in the garden of the Smith Art Gallery and Museum in Stirling, Scotland.

The harmonious blend of stone cottages and walls make Cotswold villages famously photogenic.

Further reading

All the following are available from the Dry Stone Walling Association (details opposite).

British Trust for Conservation Volunteers. *Dry Stone Walling*. 1999. Probably the most comprehensive book on the subject.

Dry Stone Walling Association. *Dry Stone Walls – The National Collection.* 2002. A record of the millennium project described opposite.

Dry Stone Walling Association. *Dry Stone Walling Techniques & Traditions.* 2004. An informative and practical guide to walling.

Griffiths, David. *In There Somewhere.* 1999. Very informative on most aspects of the craft.

Rainsford-Hannay, Colonel F. *Dry Stone Walling*. First published 1947, this is a readable pioneering survey (not a technical manual).

Tufnell, Richard. *Building and Repairing Dry Stone Walls*. A beginner's guide.

How to Build and Repair Dry Stone Walls is a popular instructional video produced by the Dry Stone Walling Association.